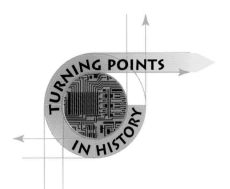

TURNING POINTS IN HISTORY

The Invention of the Silicon Chip

A Revolution in Daily Life

WINDSOR CHORLTON

Heinemann LIBRARY

www.heinemann.co.uk/library
Visit our website to find out more information about Heinemann Library books.

To order:
 Phone 44 (0) 1865 888066
Send a fax to 44 (0) 1865 314091
 Visit the Heinemann Bookshop at www.heinemann.co.uk/library to browse our catalogue and order online.

First published in Great Britain by Heinemann Library, Halley Court, Jordan Hill, Oxford OX2 8EJ, a division of Reed Educational and Professional Publishing Ltd. Heinemann is a registered trademark of Reed Educational & Professional Publishing Ltd.

OXFORD MELBOURNE AUCKLAND JOHANNESBURG BLANTYRE
GABORONE IBADAN PORTSMOUTH NH (USA) CHICAGO

Produced for Heinemann Library by Discovery Books Limited
Designed by Sabine Beaupré
Originated by Ambassador Litho Limited
Printed in Hong Kong

06 05 04 03 02
10 9 8 7 6 5 4 3 2 1

ISBN 0 431 06938 7

British Library Cataloguing in Publication Data

Chorlton, Windsor
The invention of the silicon chip: a revolution in daily life. - (Turning points in history)
1. Technology - Social aspects - Juvenile literature
I. Title
303.4'83

Acknowledgements
The Publishers would like to thank the following for permission to reproduce photographs:
BBC Worldwide, p. 19; *Corbis*, pp. 4, 5, 6, 8, 9, 10, 12, 13, 14, 15, 17, 20, 21, 24, 26, 28, 29; *Discovery Photo Library*, pp. 18 (Devendra Shrikhande), 23 (Devendra Shrikhande); *Hulton Getty*, pp. 7, 11; *NASA*, p. 23; *NOAA/National Weather Service*, p. 25; *Safeway*, p. 16; *Volvo Cars of North America*, p. 27.

Cover photographs reproduced with permission of: (top) easyEverything (bottom) *Corbis*.

Every effort has been made to contact copyright holders of any material reproduced in this book. Any omissions will be rectified in subsequent printings if notice is given to the Publisher.

Contents

Kilby's idea 4

Home life before the chip 6

Out and about 8

The first step 10

Chips and wafers 12

Rockets and calculators 14

The information revolution begins 16

Entertaining chips 18

Working with new technology 20

The shrinking world 22

Helping to make life better 24

Chips in daily life 26

The revolution continues 28

Time-line 30

Glossary 31

Index 32

Any words appearing in the text in bold, **like this**, are explained in the Glossary.

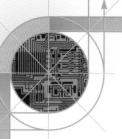

Kilby's idea

A new type of circuit

In July 1958, the Texas Instruments **electronics** plant in Dallas, Texas, closed for an annual holiday. The only engineer left at work was Jack Kilby, who had joined the company two months earlier and wasn't entitled to take time off. Kilby had been hired to develop a new type of **electrical circuit**. He wasn't happy with the proposed design because he thought it was too big and expensive.

Alone in his laboratory, Kilby tried to come up with a better design. He read and thought and sketched. Suddenly the answer came to him. Instead of making the circuit out of lots of separate parts, why not make the whole thing out of a single piece of **silicon**? Within two months, Kilby had a working model: a complete electrical circuit only 1 centimetre long. Texas Instruments called it the 'solid circuit'. Today, it is better known as the integrated circuit, the silicon chip, or the microchip.

Silicon chips are usually smaller than 5 cm along each side and sometimes no more than a millimetre. Silicon chips consist of hundreds, or even millions, of parts formed from one piece of silicon.

Before the invention of the silicon chip, there were only a few computers. The first general-purpose computer, the Electronic Numerical Integrator and Calculator (ENIAC), weighed 30 tonnes and used enough electricity to power fifteen large homes. Yet it had less computing power than a modern laptop!

The beginning of a revolution

Kilby thought his invention would make radios and televisions smaller and cheaper. He never imagined that it would transform room-sized computers into today's laptops or lead to many other inventions, from mobile telephones and barcode scanners to video games and the Internet.

Chips for many purposes

The fifteen billion chips now in existence are the building blocks of our age. They control labour-saving devices such as **robots** and photocopiers. They enable telephone systems to handle millions of calls at one time. Installed in machines such as electron microscopes or **satellite** cameras, they reveal new information about our world and beyond. As the 'brains' in computers, silicon chips can store and exchange massive amounts of information.

INVENTING THE FUTURE

In 1998, 40 years after Kilby invented the silicon chip, Tom Engibous, chairman of Texas Instruments, said, '*Jack* [Kilby] *did more than invent the integrated circuit that day – he invented the future.*'

Home life before the chip

The age of change

The 1950s was a period of rapid change. Jet aeroplanes were crossing the Atlantic Ocean and the first **satellites** were put into orbit around Earth. But people still listened to music by playing records on scratchy-sounding gramophones, as they had done 50 years before. And children often played with home-made wooden toys, as their grandparents had done.

Television viewers in the 1950s had much less choice than they have today. Early programmes were in black and white. Although there were colour televisions by the 1950s, few people could afford to buy them.

Radio and television

In the 1950s, people relied on the radio more than they do today for news and their favourite programmes. But for many, television was the **electronic** marvel of the age. By 1959, television sets in the home were common. Watching television became a popular leisure activity.

Appliances such as radios and televisions in the 1950s used **vacuum tubes**. These acted as **valves** to make electrical current flow in the right direction. They were also **amplifiers** that increased the current's strength. But vacuum tubes were fragile, used a lot of power and got very hot. In those days, radios and televisions were as large as pieces of furniture because vacuum tubes took up a lot of space. Televisions were expensive, took a long time to start up when they were turned on, and the sound and picture quality were poor. Most of all, like many appliances of the time, they often broke down completely!

Domestic appliances

Televisions became more affordable as the 1950s progressed, and so did other electrical appliances. But they were still considered luxury items. Dishwashers were unusual, and washing machines were simple machines with no electronic parts. Many people still washed their laundry by hand. There were fewer labour-saving devices in the kitchen to chop, mix and prepare foods.

Before the **silicon** chip, people spent more time on domestic chores. Most women with families worked at home rather than going out to work.

Out and about

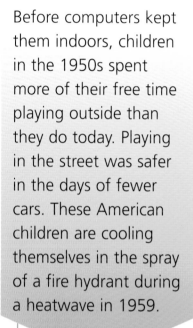

School and work

At school in the 1950s, there were no videos or computers, and so all information came out of schoolbooks. All work was written out by hand, and there were no calculators to help with sums. Instead, children used **slide-rules** and printed tables to help with their calculations.

In the 1950s, many students finished school at the age of sixteen and went to work. Most men still worked in manual jobs in the 1950s. In the country, this could be farm work, but most lived in cities or towns and worked in factories.

Women who worked outside the home often had jobs in shops or as typists, clerks or secretaries. Office workers spent a lot of time filing papers and copying documents by hand or on typewriters. There were no computers to correct typing mistakes: work just had to be done again from the beginning until it was right.

Before computers kept them indoors, children in the 1950s spent more of their free time playing outside than they do today. Playing in the street was safer in the days of fewer cars. These American children are cooling themselves in the spray of a fire hydrant during a heatwave in 1959.

Travelling and communication

In 1950 there were only a few cars on the road compared to today. Many more people made their daily journeys on trains and buses than they do now. Cars were expensive because they were largely assembled by hand. They were also much simpler, with few **electronic** parts. So, although cars broke down a lot, they were fairly easy to repair because they didn't contain lots of complicated equipment.

Workers inspect and test typewriters in an office equipment factory in the 1950s. Before the **silicon** chip, factories relied more on people to run the machinery and assemble parts by hand.

In the 1950s, people telephoning even a short distance away might need an operator to connect their calls. It could take several minutes for a connection to be made. Quite often, callers couldn't get through at all. Long-distance calls were only made on important occasions. In the 1950s, a call from London to New York cost much more than it does today. Most long-distance communication was by post. If it was urgent, people sent telegrams. These were messages sent along wires using electrical **signals**. The messages were then written out and delivered by a post office messenger.

GOING SHOPPING

Most people shopped at small, local stores. People usually walked to the shops unless they lived in the country. Supermarkets were rare except in the USA, and they weren't nearly as big as they are now. Before the silicon chip, shopkeepers had to keep track of their stock and do their ordering by making lists of what they had. The only way to do this was to count every item on the shelves.

The first step

Replacing the vacuum tube

The **vacuum tubes** in early **appliances** were large and unreliable. So scientists worked to find better ways of controlling **electronic** equipment. At Bell Telephone Laboratories in New Jersey in the USA, researchers tried making miniature replacements for vacuum tubes out of **silicon** and **germanium**. The researchers found that they could control the way electric current flowed through silicon and germanium by coating them with certain chemicals.

The invention of the transistor

Scientists Walter Brattain and John Bardeen made a breakthrough in December 1947. They applied an electric current to gold wires on a plastic triangle held over a crystal of germanium. They found that the germanium **amplified** the **signal**, making it nearly 100 times stronger. This device came to be called a **transistor**. Transistors, like vacuum tubes, could control electric current, but they were more reliable and used less power.

A few weeks after Brattain and Bardeen's discovery, their colleague William Shockley designed a more efficient type of transistor. This consisted of three layers of germanium sandwiched together. These transistors were

John Bardeen (left), William Shockley (middle) and Walter Brattain (right) are pictured here at the Bell Laboratories in 1948. They won the Nobel Prize in 1956 for their invention of the transistor.

so small that it was possible to fit the entire **electrical circuit** of a radio into the space taken up by a single vacuum tube! After 1960, transistors were made from silicon, which is cheaper than germanium and works better at high temperatures.

Transistors catch on

Transistors soon found several uses. They made telephone exchanges more efficient. In 1952, the first hearing aids to use transistors instead of vacuum tubes were produced. And the armed forces in the USA now had smaller and more reliable electronic parts to control computers and **guided missiles**.

Transistor radios first appeared in 1954. This picture from the 1950s shows how small they were compared to radios that used vacuum tubes. Transistor radios were popular with young people because they were small and cheap. Soon, radio stations began playing pop music for this new audience.

WHAT IS SILICON?

Silicon is a very common **element** found in minerals such as quartz and sand. Silicon (and germanium) are called semiconductors because they are neither pure **conductors**, like metals, nor pure **insulators**, like rubber. On a piece of silicon, one area can be chemically treated to conduct electricity, while another area can be treated to prevent the flow of electricity.

Chips and wafers

Trouble with transistors

Early **transistors** weren't perfect. They had to be made by hand, and technicians had to look through microscopes to attach tiny wires. A speck of dirt could ruin a transistor. In any batch made, it was lucky if more than half worked properly.

The poor reliability of the transistors worried the US armed forces. They found that many of their multi-million-dollar **guided missiles** failed because of faulty transistors worth only a few dollars each. And although transistors were much smaller than **vacuum tubes**, the transistorized missile guidance systems still took up a lot of space.

Jack Kilby, surrounded by some of the devices that have resulted from the invention of the silicon chip.

The silicon chip

The chip that Jack Kilby invented in 1958 solved the problems of transistors, and did much more. This miniature marvel contained tiny transistors and all the other parts of an **electrical circuit** on a piece of **silicon** no bigger than a thumbnail. The only drawback was that it was difficult and expensive to make.

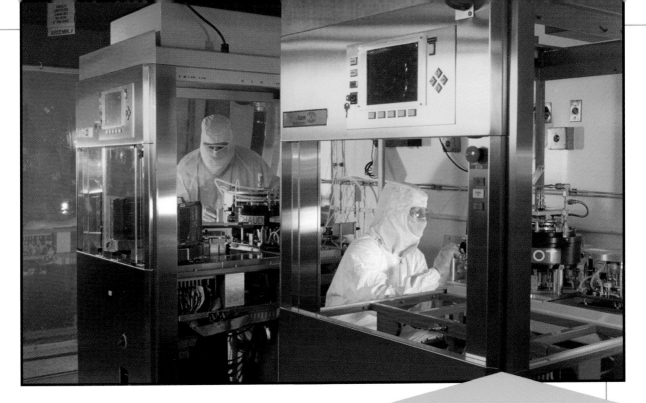

Noyce's wafers

However, Kilby wasn't the only engineer working on the problem. In 1959, at Fairchild Semiconductor in California, Robert Noyce developed a silicon chip built up in thin slices, called wafers, that could hold even more transistors in a tiny space. Kilby was awarded the Nobel Prize for his invention, but it was Noyce's design for a miniature electrical circuit that went into **mass production**.

Even the tiniest speck of dust can spoil a delicate silicon chip, so chips are made in spotless areas known as 'clean rooms'. A pattern is made on the silicon chip to form a circuit of transistors and other parts. Tiny pieces of aluminium are then added to connect the circuit together.

HOW SILICON CHIPS WORK

Like the nerve cells in our bodies, silicon chips control all the other parts. When current is applied to the circuit on a chip, the chip will direct an **electronic** device. The transistors and other parts on the chip act as **valves** and pumps, guiding the flow of the electrical current. By arranging the same parts in different ways, chips can be **programmed** to control the performance of a huge range of devices, from computers and **digital** cameras to **robots** and TV remote controls. Many devices can be linked to make complex systems, such as a telephone network or an air traffic control system.

Rockets and calculators

Chips in space

The immediate effect of the **silicon** chip was hard to see, either in business or in daily life. When silicon chips first went on sale in 1961, they cost hundreds of pounds each. Most manufacturers went on using old-fashioned **transistors**. Early chips were used mainly in **guided missiles** and spacecraft.

Costly calculators

Texas Instruments was very worried about the poor sales of its silicon chips. The company asked Jack Kilby to design a portable calculator just to show the benefits of miniature chip **technology**. Existing transistorized calculators weighed 25 kilograms. The model that Kilby produced in 1967 could fit into a coat pocket but cost hundreds of pounds. Gradually, as more chips were made they got cheaper; and so did calculators.

It's hard to believe there were **satellites** in space before there were pocket calculators. In 1958, these army scientists were launching the USA's first satellite. Amazingly, they were using a **slide-rule** (seen lying on the left on the desk) to make their calculations during the countdown.

Quartz watches were among the few products that did use silicon chips in the early days. The first quartz watches were expensive, but now they are so cheap to make that some businesses give them away as gifts.

No future for computers

The computer industry was the other main market for silicon chips. But although chips made it possible to build computers a fraction of the size of earlier models, nobody thought there would be a demand for computers in the home or office.

In 1971, Intel, a company co-founded by Robert Noyce, worked out how to put the operating parts of a computer on to a single silicon chip. Using this **microprocessor**, it was possible to build computers small enough to fit on a desk. Even then, the big manufacturers didn't believe that the general public would buy computers. Instead, microprocessors were used in household items such as televisions and microwave ovens to make them run more efficiently.

MULTIPLYING TRANSISTORS

Scientists are always working on making chips more powerful. The number of transistors that can be placed on a silicon chip doubles roughly every eighteen months. The most advanced chips of today contain 30 million transistors.

The Apollo spacecraft that landed the first men on the Moon in 1969 was guided by an onboard computer containing 5000 silicon chips. By the time of the last Apollo mission in 1975, a pocket calculator carried by one of the astronauts had more processing power than the 1969 computer.

The information revolution begins

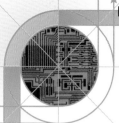

Information in barcodes

One of the biggest changes brought about by the **silicon** chip was how it enabled machines to process information. One way of doing this is with barcodes – patterns of stripes that contain information – and the chip-controlled scanners that read them. On 26 June 1974, the first barcode scanner was used at a supermarket in Troy, Ohio, in the USA. First out of the trolley was a pack of chewing gum. From then on, barcodes and scanners began appearing everywhere.

Nowadays, nearly everything you buy has a barcode. The one for this book is shown above and on the back cover. Most shops have a scanner system that reads prices from barcodes. At this supermarket, groceries pass over the scanner panel in front of the check-out clerk.

Barcodes can show prices, or how many of an item are left in stock. They can track parcels and letters all over the world. Barcodes also identify patients in hospitals. Scientists have even put barcodes on insects to monitor their behaviour.

The first desktop computers

Computers that fit on top of a desk, called personal computers or PCs, appeared in 1975. The first ones were sold as a kit to be assembled by the buyer. In 1976, Steve Jobs and Steve Wozniak started selling a pre-assembled computer, 'Apple I', which they built in a garage in Silicon Valley, California. But it was a few years before big business took notice. IBM, the

world's largest maker of business machines, finally realized that small computers could be useful. In 1981, IBM launched the IBM-PC for office use.

Computers join the workforce

Many people did not like the new **technology** at first. They worried about losing information if they pressed the wrong key. Some managers, used to having secretaries type their paperwork, resented learning to use a computer. The benefits of personal computers soon became obvious, though. With computers, it was possible to correct spelling mistakes and move text around without retyping. Most importantly, information stored and processed in computers could be found or used at the touch of a button.

Silicon Valley in the USA has been the base of the computer industry ever since the invention of the silicon chip. In the 1990s, when this picture was taken, it also became the headquarters of many hundreds of new businesses that sprang up as the Internet boom began.

SILICON VALLEY

Until the 1960s, the strip of land between San Francisco and San Jose in California was mainly a fruit-growing region. But after Noyce's invention of the silicon wafer chip, the first silicon chips were **mass-produced** there. The name 'Silicon Valley' was first applied to the area in 1971. Today, the valley is home to more than 3000 high-tech companies, including Intel, Netscape, Sun Microsystems, Oracle and Silicon Graphics.

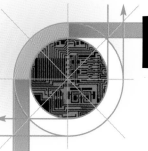

Entertaining chips

Computers at home

Soon, people began buying computers for home use. They used them for writing letters, keeping family budgets and playing games. From the mid-1980s, computer sales soared. Less than 20 years after the big manufacturers had insisted that nobody would want a personal computer, people were finding that they couldn't live without them.

Today, all kinds of things are produced at home on a computer that before would have taken a professional printer. People make their own posters and leaflets, illustrate reports and even publish magazines.

The 1980s saw the arrival of small, portable video games that could be used anywhere. New electronic games and toys appear in the shops every year. These hand-held units cost just a few pounds each and offer over 250 different games.

Video games

In 1972, American computer scientist Nolan Bushnell invented a simple bat-and-ball game that could be played on machines in bars and arcades. Bushnell called it 'Pong', after the sound made when a player hit the ball. The first machine was installed in a bar in California. The constant 'pong' noise began to attract curiosity. The next morning, people were lining up to play the new game.

The computer and video game craze had started. Within a few years, the **electronic** bleeps of computer games would become a familiar sound all around the world.

The television generation

The changes in television since the **silicon** chip have been dramatic. **Digital** images create sharp pictures, and advances in manufacturing have produced reliable, cheap television sets. Homes around the world can receive hundreds of channels beamed by **satellites** from other countries.

Special effects and **animation** on television and in movies have greatly improved. In the 1939 film *The Wizard of Oz*, a tornado was simulated by someone twirling a piece of fabric. In the 1996 blockbuster *Twister*, the scenes of tornadoes hurling trucks through the air were created using computers.

Stunning special effects are created with the help of silicon chips. In a 2000 television documentary about dinosaurs, computer animation was used to create images that appeared as if they had been filmed in real life.

THE DIGITAL AGE

In digital equipment, silicon chips convert **signals** of any kind – such as musical notes, words, or pictures – into numbers, then convert them back into their original form. This gets rid of any bad quality in the signal, which explains why digital sound recordings, such as those on compact discs, are clearer than non-digital ones, such as those on old vinyl records.

Working with new technology

Changes in the workplace

Although the **silicon** chip took many years to find its way into the workplace, there is no question that a huge change has now taken place. The kind of work people do and how they do it has altered greatly.

Robots have taken over many manual jobs such as welding, paint spraying or inspecting parts for faults. They can perform the same task perfectly over and over again. Cars, such as the ones being made in this factory, can be produced much more cheaply and much faster than before the silicon chip.

Tools and **technology** that we take for granted, and that make our jobs easier and faster, just didn't exist 20 years ago.

Chips take over old jobs

Computers are used at every stage of the business process. People use them to design offices, factories and new products. Silicon chips control the movement of products from the factory to the places where they are sold. Above all, chips control machines that have taken over from people in factories.

ROBOTS AT WORK

The first factory **robots** appeared in 1961. As robots get increasingly 'intelligent', they are used to replace people in more and more jobs, and to bring additional skills to many areas of work. But this doesn't always mean fewer jobs. As robots and technology make businesses more efficient, the companies can grow and employ more people.

Chips create new jobs

Because of this, there are fewer manual manufacturing jobs, and more people work in offices than ever before. Silicon chip technology has greatly reduced the amount of time office workers spend on routine tasks such as filing and copying. Information that would have taken people days to process before the silicon chip is now processed by a computer in minutes. But new jobs are created all the time. Computer **programming**, providing services to high-technology businesses, and working in many new forms of communication – these are the jobs of the 'computer age'.

Some robots are used to carry out dirty or dangerous tasks, such as investigating suspected bombs or moving nuclear waste. This underwater robot is picking up a piece of decorative glass from the shipwrecked *Titanic*. The ship sank in 1912, and it was years before new technology made it possible to recover some of its remains.

The shrinking world

Getting connected

Another huge change brought about by **silicon** chip **technology** is in communications. There have been many improvements in the last 40 years. It is now much easier and faster to make contact with people in different parts of the world.

In the 1950s, making an international telephone call was complicated, and a letter could take weeks to get from one country to another. Now, because telephone systems use silicon chips, millions of calls are made every day cheaply and efficiently. Telephone lines are also used for sending documents instantly on fax machines. Chips also control **satellites** and mobile telephones. With a mobile satellite phone, which sends **signals** via a satellite high above the Earth, you could call home even from the top of Mount Everest.

Satellites do a lot more than just send telephone signals. As they circle the Earth, they receive and transmit television programmes from one part of the world to another and send back pictures from space.

The Internet

The Internet wouldn't exist without silicon chips. Based on an American military computer network, the Internet started in the late 1960s. Today, any computer user with a **modem** and telephone can use the Internet. Once 'online', people can send e-mail (**electronic** mail) and documents instantly to other Internet users anywhere in the world. They can also look through masses of information on the thousands of websites that make up the World Wide Web. People use the Internet to research homework, get sports results, listen to music, watch videos and go shopping.

No need to travel

Because of all these developments in communication, there are more and more links between people and businesses in different countries. Workers thousands of miles apart can 'teleconference', using computers and video cameras to discuss ideas or show new products without leaving their offices.

Chip technology gives people the ability to go shopping anywhere in the world without leaving their homes. Internet websites offer a range of goods not found in any one city or country. And with a credit card, shoppers can pay in any currency because computers automatically convert the payment.

Helping to make life better

Medical breakthroughs

Many people live healthier and better-quality lives thanks to **silicon** chips. Computerized life-support systems monitor the condition of patients in intensive care units. Scanners controlled by chips can show detailed images of babies inside their mothers and help doctors detect any problems. If a patient with a rare type of blood needs a blood transfusion, computer **databases** enable doctors quickly to find the right type of blood anywhere in the country.

Three-dimensional computer **animation** is used to train surgeons in tricky operations where a slip of

Here you can see a patient entering a tube for an MRI (magnetic resonance imaging) scan. The computer monitor in front shows a cross-section of the patient's body as she moves through the scanner, creating a picture that can be used to detect problems inside the body.

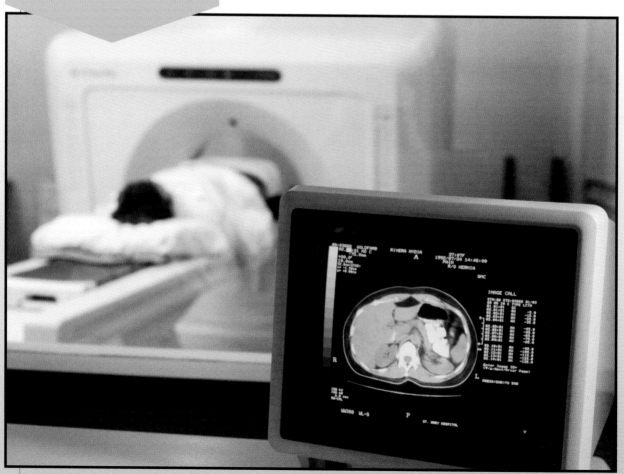

the hand could harm the patient. In the operating theatre, some types of surgical equipment are guided by **robots**, which don't suffer from unsteady hands!

Overcoming disabilities

People with heart conditions can be fitted with pacemakers. These little devices are controlled by tiny chips, and keep human hearts going in a nearly natural manner. Silicon chips implanted in the ear help deaf people hear. Blind people can 'read' with the help of a computerized machine. The device scans printed pages and converts the information into **digital** form, which in turn is translated into spoken words.

Watching the weather

Knowing what the weather will do is important. In parts of the world threatened by tornadoes or flooding, accurate forecasts can save lives. Supercomputers, which work hundreds of thousands of times faster than personal computers, have greatly improved weather forecasts because they analyze information and make predictions quickly and accurately. Fifty years ago, weather forecasts were wrong as often as they were right. Today, weather forecasters get it right nine times out of ten and can give reasonably accurate forecasts for up to five days ahead.

These weather forecasters are issuing warnings from the weather centre in Norman, Oklahoma, in the USA during a severe tornado outbreak in May 1999. With the help of their equipment and computers, they were able to issue warnings of where the tornadoes would strike. The warnings saved hundreds of lives.

Chips in daily life

Around the home

If you look at the various electrical devices in your home, you would probably find that they are all regulated by **silicon** chips. Cookers, microwaves, dishwashers and washing machines have all become cheaper to produce and more reliable to use. Because of this, many more people now own labour-saving devices. Household chores take up much less of our time than they used to.

As more and more computers appear in schools, it becomes easier to find information from anywhere in the world by using the Internet.

Chips in the classroom

Chip **technology** has opened up a new world of learning. The contents of a large, expensive set of encyclopaedias costing hundreds of pounds can now be fitted onto a cheap CD-ROM. **Animation** and sound can be added, and the tiny package becomes a huge source of reference on a classroom computer. Most schools use chip technology for everything from their science equipment to exchanging information with other schools.

Chips on the road

The use of computers and **robots** in factories has made cars more affordable and dependable, and the vehicles themselves use more and more silicon chip technology to operate. The onboard computers fitted to many new cars monitor how much fuel they use and if they are giving off polluting fumes. Some dashboard displays let the driver know when their car needs servicing, or even if a tyre needs more air in it.

This navigation system on a car dashboard can be **programmed** to help drivers find their way. The driver enters his or her destination and location, and the car's computer maps a route for the driver to follow. Some cars now have a global positioning system that uses **satellite signals** to show drivers exactly where they are.

The revolution continues

How new inventions change lives

All important inventions cause **revolutions**. The invention of the steam engine 200 years ago speeded up transport and made **mass-produced** goods affordable for ordinary people. But it also produced awful factory working conditions and terrible slums.

The invention of the **silicon** chip means that even the smallest, cheapest hand-held device can process a huge amount of information compared to the largest, most expensive pre-chip computer. But we are really only just beginning to realize how this revolution has affected work and daily life. As we have seen, in many ways the chip has made our lives safer, more comfortable and more entertaining. In other ways it has made our lives less active and more impersonal.

These astronauts from different countries are living together in the International Space Station that orbits Earth. Silicon chip technology has made life in space possible, and is helping break down barriers between nations.

The world in a screen

The television and computer screen have become a window on the world for many people. But at what price? People who spend hours in front of screens for work, and then hours more for leisure, spend little time on physical activity and become less healthy as a result.

Silicon chip **technology** has shortened distances and helped to break down national barriers because there is more communication between different parts of the world. People living thousands of miles apart can work together without ever meeting face to face or going to an office. Business and shopping can be done online. Students can complete university without ever setting foot in a lecture hall.

Thanks to the silicon chip, there is also plenty of entertainment to be had without ever leaving the house. But what happens to people who no longer need to gather together for work and play? It may be that they lose certain personal skills, or become cut off from real life.

Scientists predict that, within this century, computers will be more 'intelligent' than people. It is up to us to decide whether the machines of the future become our servants or our masters.

This electronic housedog was shown at a 'House of the Future' exhibition in 2000. It is remote-controlled, obedient and clean. The exhibition organizers believe that **robot** pets will soon be popular alternatives to the real thing.

Time-line

1945	First general-purpose electronic computer (ENIAC) goes into service
1947	John Bardeen, Walter Brattain and William Shockley invent the transistor
1954	First transistor radios go on sale
1958	Jack Kilby invents the silicon chip
1959	Robert Noyce invents the silicon wafer chip
1961	First silicon chips go on sale
1967	Jack Kilby designs one of the first portable calculators
1969	Silicon chips used in the onboard guidance system of Apollo spacecraft
	US Defense Department develops forerunner of the Internet
1971	Intel produces the first microprocessor
1972	First commercial video game, 'Pong', invented by Nolan Bushnell
1974	First barcode scanner used
1975	Sony produces home videotape system
	First personal computers sold in kit form
1976	Launch of Apple I computer
1977	Launch of Apple II computer
1979	Apple produces the first commercially successful home computer
	Cellular telephones invented
1981	IBM produces a personal computer for office use
1983	First compact discs (CDs) on sale
1985	Windows computer operating system invented by Microsoft
1990	World Wide Web created
1995	Digital Video Disc (DVD) invented
	Toy Story, the first film using entirely computer-generated imagery, is released

lossary

amplifier	electronic device that increases the strength of a signal, such as a radio signal
animation	appearance of movement, as in cartoons
appliance	device that performs a task
conductor	substance used to carry electricity
database	collection of information for processing by a computer
digital	coded as numbers
electrical circuit	arrangement of components through which a current can flow
electronic	relating to electrons, the basic particles of electricity, and used to describe devices that operate using electronic power
element	one of about 100 simple substances that make up other substances
germanium	element used for making transistors in the 1950s
guided missile	weapon that travels through the air or water directed by remote control
insulator	substance used to stop the flow of electricity
mass production	manufacture of large numbers of products by machines
microprocessor	main operating parts of a computer when placed on a single silicon chip
modem	device that enables computers to send and receive data through a telephone line
program	to give instructions to a computer or other electronic device by putting in information
revolution	important change in the way things are done
robot	computer-controlled device programmed to do work
satellite	something that goes around the Earth in space
signal	electrical current used to transmit sound or other information
silicon	element used for making electrical circuits
slide-rule	measuring device used for making calculations
technology	knowledge and ability that improves ways of doing practical things
transistor	small electronic device that controls the direction of electrical current or acts as an amplifier
vacuum tube	tube from which air has been removed that was used as an amplifier in early electronic devices
valve	device that controls flow of liquid, gas or electricity through an opening

Index

aeroplanes 6
amplifiers 6-7, 10
animation 19, 24, 26
appliances 6, 7, 15, 26

barcodes 16
Bardeen, John 10
Brattain, Walter 10
Bushnell, Nolan 18

calculators 8, 14, 15
cameras 5, 13
cars 8, 9, 20, 27
communications 22-23, 29
computers 5, 8, 11, 13, 14,
    15, 16-17, 18, 19, 20, 21,
    23, 24, 25, 26, 27, 28, 29

electrical circuits 4, 11,
    12, 13
Engibous, Tom 5
ENIAC 5

factories 9, 20, 27, 28
Fairchild Semiconductor 13

germanium 10, 11
guided missiles 11, 12, 14

health 24, 28

IBM 16-17
Intel 15, 17
Internet, the 5, 17, 23, 26

Jobs, Steve 16

Kilby, Jack 4, 5, 12, 13, 14

medical advances 24
microprocessors 15
music 6, 11, 19, 23

navigation 27
Nobel Prize 10, 13
Noyce, Philip 13, 15, 17

offices 8, 17, 20, 21,
    23, 29

pets 27, 29

radio 5, 6, 7, 11
robots 5, 13, 20, 21, 24,
    27, 29

satellites 5, 6, 14, 19, 22
scanners 5, 16, 24
school 8, 26-27
Shockley, William 10
shopping 9, 16, 23, 29
silicon 4, 10, 11, 12
Silicon Valley 16, 17
slide-rules 8, 14
space 14, 15, 22, 28
special effects 19
steam engine 28

telegrams 9
telephones 5, 6, 7, 9, 11,
    22, 23
television 5, 13, 15, 19,
    22, 28
Texas Instruments 4, 14
toys 6, 18
transistors 10-11, 12,
    13, 15
typewriters 8, 9

vacuum tubes 6-7, 10,
    11, 12
video games 5, 18-19

wafers 13, 17
weather forecasts 25
work 7, 8, 17, 20-21, 23,
    28, 29
Wozniak, Steve 16

Titles in the *Turning Points in History* series include:

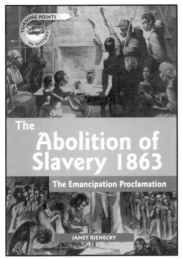

Hardback   0 431 06937 9

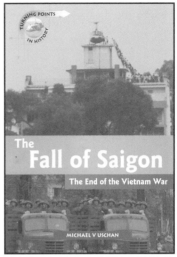

Hardback   0 431 06931 X

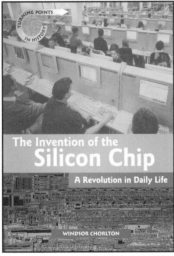

Hardback   0 431 06938 7

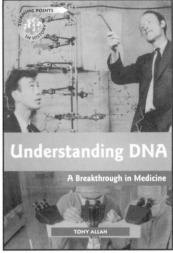

Hardback   0 431 06939 5

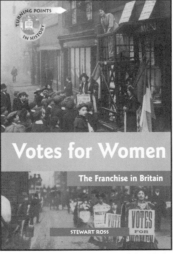

Hardback   0 431 06940 9

Find out about the other titles in this series on our website www.heinemann.co.uk/library